I0471798

Visualizing Math
Math Help

Allen Harrington

Visualizing Numbers

DEDICATION

I would like to dedicate this book to the amazing teachers I have had. I would like to thank Mrs. McLeod, Mrs. Williams, Ms Celi, Ms Gerry and many more. Thank you for taking your time and helping me comprehend all of that complex school work.

Oh and keeping me ORGANIZED!

Visualizing Numbers

CONTENTS

Visualizing Numbers

THE BEGINNING

I was a late bloomer when it came to speech and writing. When it came to talking all I did was grunt and groan until I was five and a half years old, that's when my parents were in a car accident with me in the back seat on our way home from Florida. After the car accident nobody could stop me from speaking.

When I was in Grade 4 I still used my fingers while puzzling over a math question. When my Math teacher seen this she was not pleased and sent me to the special education room as I was deemed below grade level by the Math teacher for using my fingers in math class. That night I was afraid because the only way I could do math was with my fingers and now I was told to stop.
The next day I had math again and I looked at the page and there was a number 4, then it hit me. I could picture four dots over top of the 4 like the dice in the game trouble. that's when I started thinking, I could do this for all numbers.

Since that day I have been doing math in my head visualizing each number as dice. This book will explain how this is possible and how your able to teach this with your student, child or maybe even use this method for yourself.

Visualizing Numbers

CHAPTER 1
THE NUMBERS

The only numbers you need to remember are 1 through to 9.

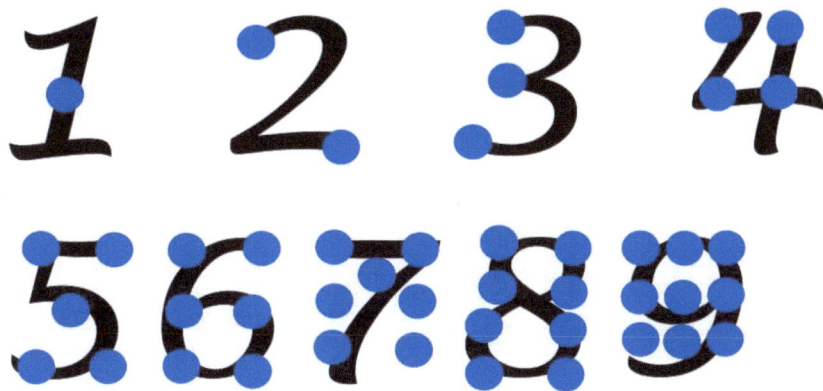

As you can see, each number has its own dot formation to easily memorize where the dots are. Below is an explanation how to memorize where the dots go in accordance to the numbers.

#1- One dot in the middle.
#2- One dot on each point of the number two.
#3- One dot on each point of the number three.
#4- Visualize four dots on the number four in the formation on dice.
#5- Visualize five dots in the formation on a dice.
#6- Visualize six dots in the formation on a dice.
#7- This number is a little harder for some people. Visualize a number six dice with a dot in the middle.
#8- With the number eight you can visualize two #4 dice. One number four dice can be visualized on the top circle of the eight and the second four dice can be visualized on the bottom circle of the eight.
#9- Visualize a 3 by 3 plot of dots over the number 9.

If you find it easier to visualize the dots in a different formation that is fine.

CHAPTER 2
ADDING

Adding with this method is fun. When adding visualized numbers you will add the dots together. Each dot has a value of one. Here's an example:

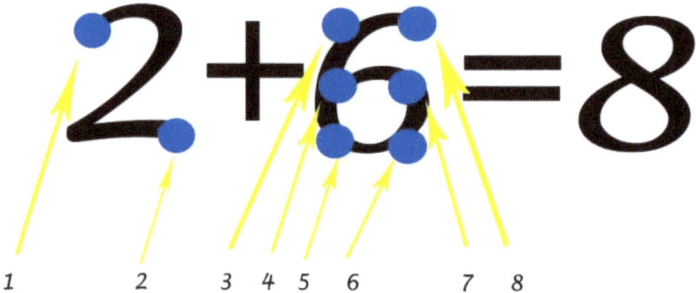

Once you become familiar with this method you can start counting the six first and add the two points on the two. When using this method always start with the biggest number in the equation, this will make calculating much faster.

Here's an example:

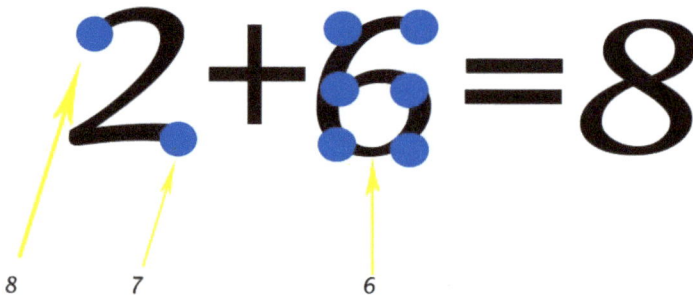

Try the following questions to perfect the method.

2+6=_____

5+6=_____

8+9=_____

$$89 \quad 56$$
$$+ \quad 3 \quad +26$$
_____

_____ _____

Now try the following questions visualizing the dots on the numbers in your mind.

3+6=_____

4+7=_____

EQUATION SHEET

Let's try some equations with the dots.

$2 + 3 =$ _____ $7 + 9 =$ _____

$5 + 9 =$ _____ $4 + 2 =$ _____

$6 + 8 =$ _____ $3 + 5 =$ _____

$$\begin{array}{r} 21 \\ +53 \\ \hline \end{array} \qquad \begin{array}{r} 55 \\ +24 \\ \hline \end{array}$$

$$\begin{array}{r} 9 \\ +85 \\ \hline \end{array} \qquad \begin{array}{r} 62 \\ +18 \\ \hline \end{array}$$

EQUATION SHEET

Now let's try the following equations while visualizing the dots on the numbers in your mind.

$$3+8=\underline{\hspace{1.5cm}} \qquad 8+2=\underline{\hspace{1.5cm}}$$

$$9+4=\underline{\hspace{1.5cm}} \qquad 6+8=\underline{\hspace{1.5cm}}$$

$$5+6=\underline{\hspace{1.5cm}} \qquad 7+4=\underline{\hspace{1.5cm}}$$

$$
\begin{array}{r}
9 \\
+\ 35 \\
\hline
\end{array}
\qquad
\begin{array}{r}
67 \\
+\ 61 \\
\hline
\end{array}
$$

$$
\begin{array}{r}
8 \\
+\ 47 \\
\hline
\end{array}
\qquad
\begin{array}{r}
99 \\
+\ 85 \\
\hline
\end{array}
$$

Chapter 3
SUBTRACTING

Subtracting when visualizing your numbers is much like those visuals of beads the teachers used to use. (My teachers always used beads as an example). For a question like four minus two the teacher would place 4 beads on the table and take two away, which would leave two beads which would bring the answer of two.

This is much the same. If the question is eight minus six, the six would eat 6 dots from the 8 which would leave 2 dots remaining. The final answer is two.

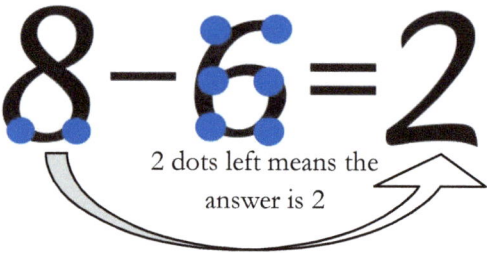

2 dots left means the
answer is 2

If there is an equation where you need to subtract the same number from itself for example, six minus six. The second six would eat the first numbers dots which would leave you with no dots and an answer of zero.

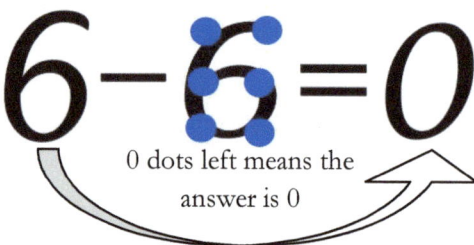

0 dots left means the
answer is 0

Let's try this method out:

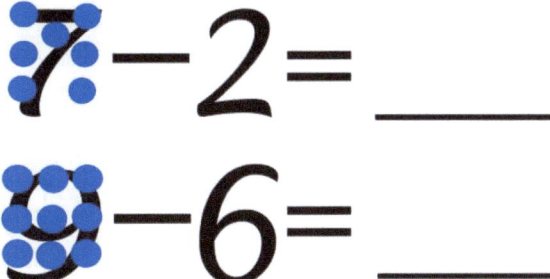

$$7-2=\ _____$$

$$9-6=\ \_\_\_\_$$

Now let's try to picture the dots in our head while doing the following equation. Remember that the second number eats its number of dots from the first number.

$$5-4=\ _____$$

$$6-2=\ _____$$

Vertical Subtraction is the exact same. The bottom number eats its number of dots from the top number. The example below shows 8 minus 8, The bottom eight ate all of the top numbers dots. The answer is zero.

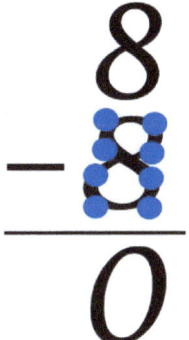

EQUATION SHEET
Let's try some equations with the dots.

$$6-5=\underline{\qquad} \qquad 2-1=\underline{\qquad}$$

$$8-7=\underline{\qquad} \qquad 6-4=\underline{\qquad}$$

$$8-4=\underline{\qquad} \qquad 9-6=\underline{\qquad}$$

$$\begin{array}{r} 6 \\ -\ 6 \\ \hline \end{array} \qquad \begin{array}{r} 99 \\ -17 \\ \hline \end{array}$$

$$\begin{array}{r} 29 \\ -11 \\ \hline \end{array} \qquad \begin{array}{r} 26 \\ -\ 7 \\ \hline \end{array}$$

EQUATION SHEET

Now let's try the following equations while visualizing the dots on the numbers in your mind.

$$8-3 = \underline{\hspace{2cm}} \qquad 8-2 = \underline{\hspace{2cm}}$$

$$9-4 = \underline{\hspace{2cm}} \qquad 8-6 = \underline{\hspace{2cm}}$$

$$6-5 = \underline{\hspace{2cm}} \qquad 7-4 = \underline{\hspace{2cm}}$$

$$\begin{array}{r} 35 \\ -\ 9 \\ \hline \end{array} \qquad \begin{array}{r} 67 \\ -61 \\ \hline \end{array}$$

$$\begin{array}{r} 8 \\ -\ 7 \\ \hline \end{array} \qquad \begin{array}{r} 99 \\ -85 \\ \hline \end{array}$$

CHAPTER 4
MULTIPLICATION

For most students multiplication is very difficult, if you follow this method they will be doing multiplication with ease.

Multiplication while visualizing your numbers is fun and fast to do. If the question is six multiplied by six you would picture the second number with its dots and each of those dots would be the same value as the first number (look below).

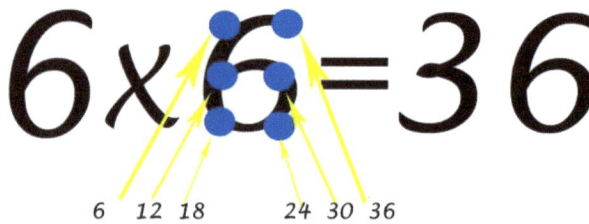

6   12   18   24   30   36

If both numbers are not the same for example, seven times three. You would then picture the dots on the three and each dot will be worth seven. If you have difficulties counting by seven, you can picture seven dots on the seven and each dot will be worth three.

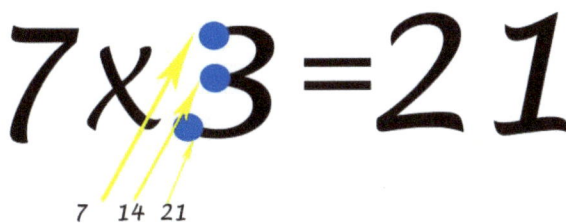

or

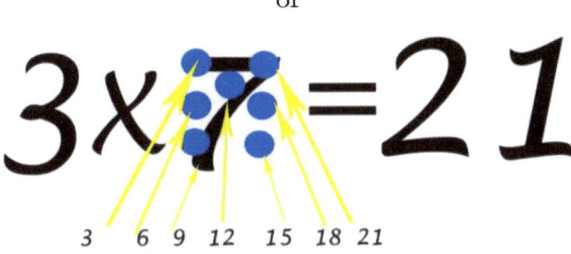

Let's try the following multiplication questions.

$$3 \times 5 = \underline{\hspace{2cm}}$$
$$4 \times 3 = \underline{\hspace{2cm}}$$

Visualize the following questions. Remember to use the easiest number you can count up by.

$$3 \times 3 = \underline{\hspace{2cm}}$$
$$4 \times 6 = \underline{\hspace{2cm}}$$

Very good, let's try vertical multiplication now. Vertical multiplication will be done with the same process. Remember starting farthest to the right of the equation.

Try Drawing your dots

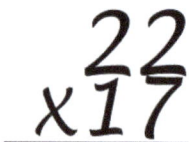

$$\begin{array}{r} 6 \\ \times\ 6 \\ \hline \end{array} \qquad \begin{array}{r} 22 \\ \times 17 \\ \hline \end{array}$$

EQUATION SHEET

Now let's try the following equations while using the dots on the numbers.

$6 \times 5 =$ _____ $2 \times 1 =$ _____

$8 \times 7 =$ _____ $6 \times 4 =$ _____

$8 \times 4 =$ _____ $9 \times 6 =$ _____

$$\begin{array}{r} 6 \\ \times\ 6 \\ \hline \end{array}$$

$$\begin{array}{r} 99 \\ \times\ 17 \\ \hline \end{array}$$

$$\begin{array}{r} 29 \\ \times\ 11 \\ \hline \end{array}$$

$$\begin{array}{r} 26 \\ \times\ 7 \\ \hline \end{array}$$

EQUATION SHEET

Now let's try the following equations while visualizing the dots on the numbers in your mind.

$7 \times 5 = \underline{\hspace{2cm}}$ $3 \times 5 = \underline{\hspace{2cm}}$

$8 \times 9 = \underline{\hspace{2cm}}$ $9 \times 3 = \underline{\hspace{2cm}}$

$4 \times 4 = \underline{\hspace{2cm}}$ $1 \times 6 = \underline{\hspace{2cm}}$

$$\begin{array}{r} 6 \\ \times\ 5 \\ \hline \end{array} \qquad \begin{array}{r} 97 \\ \times\ 13 \\ \hline \end{array}$$

$$\begin{array}{r} 18 \\ \times\ 91 \\ \hline \end{array} \qquad \begin{array}{r} 9 \\ \times\ 7 \\ \hline \end{array}$$

Advanced Math Visualizing Tips

Let's try some B.E.D.M.A.S. Math
i.e.
(2+6)x7=

Brackets
Exponents
Division
Multiplication
Subtraction

You do the same as what you have learned previously in this book however, remember with B.E.D.M.A.S. you will start the equation inside the brackets and finish with subtraction.

The next few pages are the answer keys for each of the equation sheets in this book plus a notes section

ADDING

SHEET 1: 5, 16, 14, 6, 14, 8, 74, 79, 94, 80
SHEET 2: 11, 10, 13, 14, 11, 11, 44, 128, 55, 184

SUBTRACTING

SHEET 1: 1, 1, 1, 2, 4, 3, 0, 82, 18, 19
SHEET 2: 5, 6, 5, 2, 1, 3, 26, 6, 1, 14

MULTIPLICATION

SHEET 1: 30, 2, 56, 24, 32, 59, 36, 153, 29, 62
SHEET 2: 30, 15, 72, 27, 16, 6, 30, 111, 98, 63

NOTES:

Visualizing Numbers

Allen Harrington

I hope you enjoyed Visualizing Math, this technique has helped me tremendously throughout my whole life. I hope this method can help many more.

This is my first published book (there are many more to come). For more information about myself or to read more of my books check out my website: www.alsbballtraining.ca/author

Sincerely

Allen Harrington

Visualizing Numbers

www.ingramcontent.com/pod-product-compliance
Lightning Source LLC
Chambersburg PA
CBHW041121180526
45172CB00001B/359

* 9 7 8 1 4 7 7 5 8 7 8 5 0 *